Community Breeders' Associations For Dairy Cattle Improvement

by Wisconsin Agricultural Experiment Station

with an introduction by Jackson Chambers

This work contains material that was originally published in 1910.

This publication is within the Public Domain.

This edition is reprinted for educational purposes and in accordance with all applicable Federal Laws.

Introduction Copyright 2018 by Jackson Chambers

Self Reliance Books

Get more historic titles on animal and stock breeding, gardening and old fashioned skills by visiting us at:

http://selfreliancebooks.blogspot.com/

Introduction

I am pleased to present another title in the "Cattle" series.

The work is in the Public Domain and is re-printed here in accordance with Federal Laws.

As with all reprinted books of this age that are intended to perfectly reproduce the original edition, considerable pains and effort had to be undertaken to correct fading and sometimes outright damage to existing proofs of this title. At times, this task is quite monumental, requiring an almost total "rebuilding" of some pages from digital proofs of multiple copies. Despite this, imperfections still sometimes exist in the final proof and may detract from the visual appearance of the text.

I hope you enjoy reading this book as much as I enjoyed making it available to readers again.

Jackson Chambers

DIGEST

The Purpose of Community Organization for breeders of dairy cattle is to secure the cooperation of the various breeders of a community in the production and improvement of high grade and pure-bred dairy cattle and in establishing a reputation for the community as a breeding center. The advantages of this plan are: concentrated attention along definite lines; cooperation of the breeders in establishing high standards for the community and encouragement for the owners of several herds to produce a distinct type and to supply a large number of such animals to meet the demand created by cooperative advertising. Through the meetings of the organization its members are kept informed of the progress in all that pertains to their work. Through cooperation they may secure protection against fraud and contagious diseases and may secure many advantages not available to the individual breeder. Breeders' associations can secure official tests for advanced registry at less cost than where such associations do not exist.
Pages 3-4

Community Organizations in Wisconsin were begun in 1906 in Waukesha County with the organization of the Waukesha Guernsey Breeders' Association. Since that time other organizations have been formed until in January, 1910, there were 31 such associations, representing the Guernsey, Holstein and Jersey breeds or a collection of all breeds of dairy cattle in the community. These organizations are distributed through the important dairy sections of the state and include most of the leading breeders of dairy stock.
Pages 4-7

Advantages in Buying and Selling are secured through cooperative advertising; through the purchase of best breeding males for use in several herds and through cooperation in purchasing and importing a number of choice animals. By exchanging animals the members may improve their herds with good blood without importing animals from a distance. The members may combine in selling to fill large orders for a distinct dairy type and thereby attract buyers from important dairy states. Carloads of selected animals have been shipped from Wisconsin to many middle states, the Pacific coast and even to Japan and Mexico as a result of such organization in this state.
Pages 8-13

The Method of Organization is for a leader in the community to call a meeting and get as many interested breeders as can be secured; this meeting to be addressed by a capable speaker on the value of such organization. A constitution and by-laws, such as are suggested in this bulletin, may be adopted and officers selected either at the initial meeting or at another meeting, as desired. Details of the organization may be varied according to local conditions. The cooperation of the College of Agriculture may be secured and wherever possible speakers will be furnished to address meetings held to organize such associations. This bulletin contains a list of community organizations formed since the spring of 1906, together with charter and present memberships, annual dues, increase in number of pure-bred sires since time of organization and number of pure-bred animals owned by members.
Pages 14-21

Community Breeders' Associations for Dairy Cattle Improvement

G. C. HUMPHREY.

Dairy farming has developed rapidly in Wisconsin during the past few decades. In 1907, according to the 1908 report of the Wisconsin State Board of Agriculture, there were in Wisconsin 1,143,606 cows, valued at $24,834,465, as compared with 799,104 head of other cattle valued at $9,794,883. Apparently there has been a marked increase each year in the number of cows kept in Wisconsin and it will be of interest to note the aggregate number of cows in the state in the 1910 Census. It is true, however, that the improvement of the dairy cow has not kept pace with her increased prominence. Thousands of farmers who profess to be dairymen are keeping and milking cows of inferior and unprofitable types today, the production of which is exceeded by better bred cows by at least 50 per cent. To increase the production of the cows of Wisconsin 50 per cent would be a great boom to the dairy industry and would return to dairy farmers a great reward for their efforts in making such an improvement.

The improvement of any class of farm animals as well as the maintenance of any improvement which may have been achieved necessitates strict adherence to certain fundamental principles of breeding and management. Many farmers and dairymen do not understand and realize the importance of these principles and consequently fail to profit by them. Other dairymen realize their importance and know better than to follow the methods they practice, but apparently lack conviction or possibly executive ability which would enable them to realize better returns for their labor. Not only is knowledge required for the successful management of all classes of live stock but also untiring energy and perseverance, prompted by the keenest possible interest in the end to be achieved. It is evident from the inability and failure on the part of men to improve their live stock that some organization is needed that will educate, encourage and help them to better their conditions. It is the object of this bulletin to outline and bring before the dairymen of the state a plan which is being introduced for the improvement of dairy cattle.

The Plan of Organization

The plan is not altogether new. It is one of local organization for the accomplishment of specific ends which pertain to the improvement of dairy cattle and to the general welfare of the community from the dairymen's standpoint. This plan has been tried out and the results have been very satisfactory.

The first organization in Wisconsin took definite form in the spring of 1906 when less than a dozen young men of Waukesha county drafted a constitution and by-laws and became the charter members of what was termed and is known today as the Waukesha Guernsey Breeders' Association. Since the organization of this association, 30 other associations have been organized in the state representing the Guernsey, Holstein and Jersey interests and the results achieved by each of these associations prove the feasibility of the plan. Figure 1 shows the location of the local breeders' associations in the state with reference to the respective breeds represented.

Each association has for its object the production and improvement of high grade and pure-bred dairy cattle of some specific breed and the establishment of cordial relations and co-operation between its members in the practice of such methods of care and management that will insure the most successful and economical results. It is the duty of every member to improve his herd by mating his cows exclusively with pure-bred bulls of the breed represented by his association, to care for his herd in an up-to-date manner, and to co-operate with his fellow members in the use of pure-bred bulls, in buying and selling animals and in promoting the general welfare of the dairy interests of the community.

Advantages of the Plan

The advantages of the plan are many and are secured through the organized efforts and activities of the members of each association. Organization for the accomplishment of a specific purpose concentrates and directs every energy toward

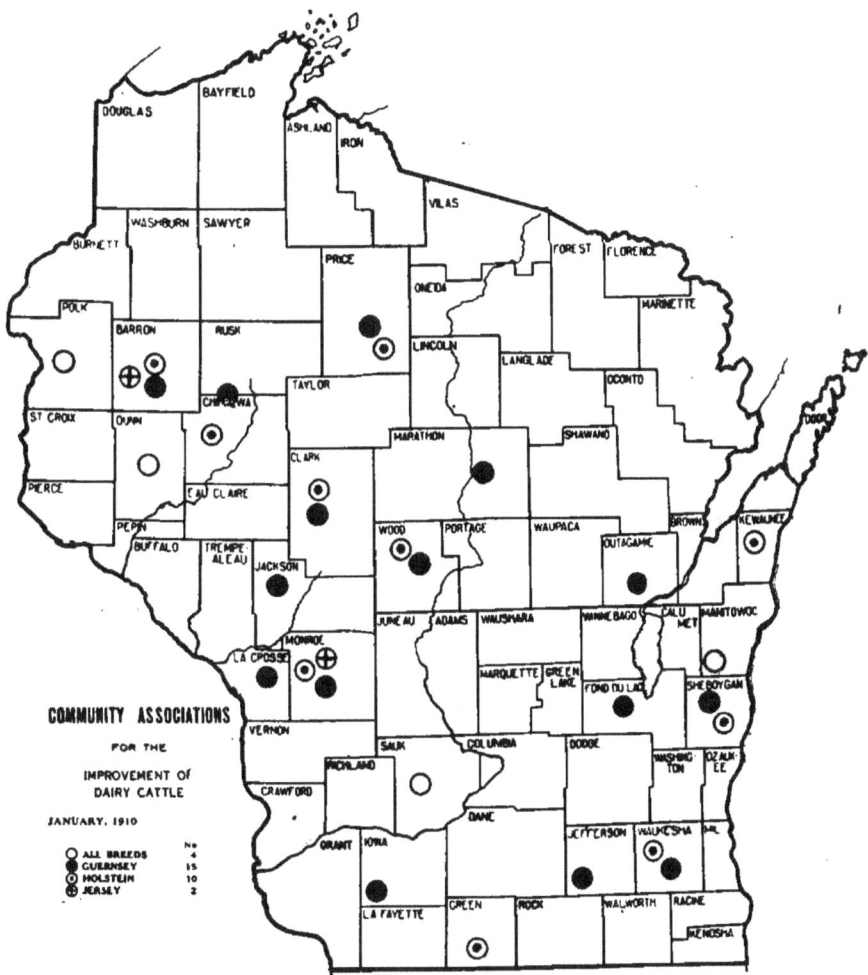

Figure 1. Location and character of community associations in Wisconsin.

that end. It tends to harmonize and center the minds of men on one project, which, if feasible, is sooner or later accomplished. No plan is more feasible than to have all the cattle of a herd and a large number, if not all of the herds in a given community of one distinct type and breed. It is apparent that there has been no organized effort in the breeding of the cattle of the great majority of herds of this state. The cattle represent promiscuous and haphazard breeding, on the whole, which, together with the same kind of care and feeding, accounts for the large number of inferior and unprofitable cows to be found in the country. Community organization for the improvement of dairy cattle creates a new interest in the subject of breeding; provides a means for the education of men in the breeding, feeding and management of their herds; elevates men out of old ruts which have seriously retarded, if not prevented their progress and places them in a position to share with their fellowmen all the opportunities which lead to success.

EDUCATIONAL ADVANTAGES.—However well educated a man may be he is always confronted with ignorance which he must continually seek to overcome. In order to pursue successfully any line of work, one must secure all the information that is obtainable and applicable to his business. He needs the constant help and counsel of his brightest and keenest associates who are engaged in the same kind of work. This is true of all dairymen and stockmen. If one allows himself to become isolated he is in danger of losing his opportunity to produce or secure the best types of animals, and of failing to feed and care for them in the right manner, or of missing a chance to sell surplus stock in a manner to insure profitable returns for the work and expense of producing it. A community breeders' association offers its members an opportunity to keep themselves informed on all that pertains to success in their work. Through the meetings of the association, men exchange helpful ideas and get the experience of prominent breeders who may be invited to address them. The experiences of many men make it possible to remedy many evils and annoyances with which dairymen have to contend.

Measures can be adopted which offer protection against fraud and unfair treatment on the part of men who might be unscrupulous in their dealings, and against the spread of contagious diseases which frequently work hardships in many communities. The state agricultural college and experiment station and the state and national dairy and live stock associations recognize the local associations as splendid opportunities for educating people and promoting the interests of cattle breeding and dairy husbandry. They are glad to assist as much as possible in providing speakers for meetings and literature for distribution, all of which is of educational value.

OFFICIAL TESTS FOR BREEDERS' ASSOCIATIONS[1]

"The organization of breeders' associations which has recently taken place in different parts of our State, will render it possible to reduce materially the cost of Advanced Registry testing for members of these associations, since the supervisors can make these tests in a circuit without loss of time and with a saving in traveling and other expenses. On account of the importance of official testing of dairy cows to the individual breeder and to the dairying industry of our state, we are anxious to do all we can to have more and more breeders take up this work, and arrangements have been made, therefore, by which tests for Advanced Registry for members of breeders' associations, and for breeders similarly situated, will be conducted at the following rates:

For 2-day tests (required by the H-F. Assn. and A. J. C. C...... $5
For 1-day tests (required by the A. G. C. C.).................... $3

provided at least six breeders in an association take up the work, so that our supervisors can make the tests on a circuit without a loss of time between the tests.

The traveling expenses of the supervisors will be charged to the association and apportioned among the members for whom tests are conducted. As in all official tests the cost of acid, notary fees, (when required) and express charges on Babcock testers, are paid by the breeders, but where six or more breeders combine these expenses will not amount to much for each party.

Under this reduced schedule of prices Jersey breeders will be able to test 10 cows at a time, and may get a majority of them into the Register of Merit, at an expense only slightly above $30 per year, since the A. J. C. C. pays half of the expenses of testing cows admitted to this Register. The cost to breeders of Holstein and Guernsey cows for the year will be about $60 and $36, respectively. Holstein breeders will have a chance to further reduce this cost by competing for prizes in the yearly tests which the H-F. Association offers, beginning with the 1910 testing year.

This arrangement places the expense of getting cows into the Advanced Registers of the leading dairy breeds within the reach of all dairymen and renders yearly testing one of the cheap-

---

[1] By F. W. Woll, in charge of dairy cow tests conducted by this Station.

est and most effective methods of advertising good stock that a breeder can adopt. The value of this system of testing to the dairy breeder in aiding him in his breeding operations will prove to be at least as great as the direct advertising he will get from the tests. Arrangements for conducting such circuit tests within local breeders' associations should be made through the secretary of the association with whom the details and appointments relative to the conduct of tests will be arranged by F. W. Woll, Madison, Wis., in charge.''

ADVANTAGES IN BUYING AND SELLING

BUYING.—If the breeders in a community are about to organize or have recently done so there will necessarily be more or less buying to do. Members will suit themselves as to how many cows they will buy, but they should pledge themselves to mate their cows with pure-bred bulls and for each member to own his own bull will undoubtedly prove most satisfactory. It is possible for the breeders to co-operate in buying and using one bull to serve several herds, but the utmost care must be used in doing this not to spread contagious abortion and other diseases of an infectious nature. If a carload of bulls or cows is desired in a certain community, it will be possible for a committee of two or three men to go and purchase what is needed at one time, which could be done at a much less cost than if each member went individually to secure what was wanted.

After a community has become established in its breeding operations, members can buy of one another without going to the expense of traveling and paying freight charges, and at the same time be much more certain of what they are getting than if they had to deal with strangers in some other county or state. In many instances it is feasible for members of an association to simply exchange bulls and thus gain a great advantage over a breeder who is isolated and not in a position to exchange bulls or to keep posted on what is in the market. The opportunity to become posted and to buy intelligently without the great risk of being deceived, as one is more or less likely to be when dealing with animals and dealers with whom he is unacquainted is one of the great advantages of community effort.

SELLING.—Likewise there are many advantages to be had through community organization in selling surplus cattle. A surplus of uniformly bred cattle of some distinct dairy breed, which it is possible for a community to produce in a few years of persistent effort, is very attractive to buyers. The demand for such cattle is very much greater than many farmers and dairymen realize. Chicago, for instance, with a population over two-thirds as great as that of the whole state of Wisconsin creates a great demand for dairy cows through the vast amount of milk and dairy products, which its population consumes. Milwaukee, Saint Paul, Minneapolis, Duluth and Superior with rapidly increasing populations, to say nothing about the smaller cities and towns in the state, use an enormous amount of dairy products. Very little attempt is made to breed and rear cows in the vicinity of the large cities. Dairymen in such sections pay liberal prices for high grade dairy cows whenever they need to increase the production of their herds or replace cows that are no longer profitable milk producers. Cows for such sections are usually purchased in carload lots by expert buyers who appreciate the advantage of being able to secure a carload of cows of a distinct dairy type in a single community. It involves a much less expenditure of time and money and much less difficulty in assembling cows for shipment than is incurred where two or more counties must be searched in order to find a carload.

A remarkable increase in the demand for dairy cattle from the states of the Middle West and Pacific Coast has been noted during the past few years. Japan and Mexico have taken several carloads of dairy cattle from Wisconsin at prices which have induced many breeders to part with their very best animals. Thus far the demand has been much greater than the supply and it is hoped through the organized effort by communities that more of this demand for dairy cattle can be met. An organized community can much better satisfy and handle the needs of parties from a distance who buy cattle in carload lots than can any one individual. Very few individuals are in a position to sell a carload of dairy cattle at one time. Well organized communities are more apt to gain recognition and attract buyers from all parts of the world than are individual men.

The prominence of Guernsey Island for Guernsey cattle, the Island of Jersey for Jersey cattle, and the little district of Holland for Holstein cattle are splendid examples of how a small section of territory may become recognized as the source of a given breed of cattle. It is quite safe to say that these two islands and the small district of Holland would not have been so conspicuous or especially prominent in the breeding world, if it had not been for their organized efforts in the breeding of cattle.

The small town of Lake Mills, Wisconsin, is another example of what community effort can accomplish in the way of breeding one specific kind of cattle. Through the efforts of a few men many Holstein herds exist in the vicinity of Lake Mills.

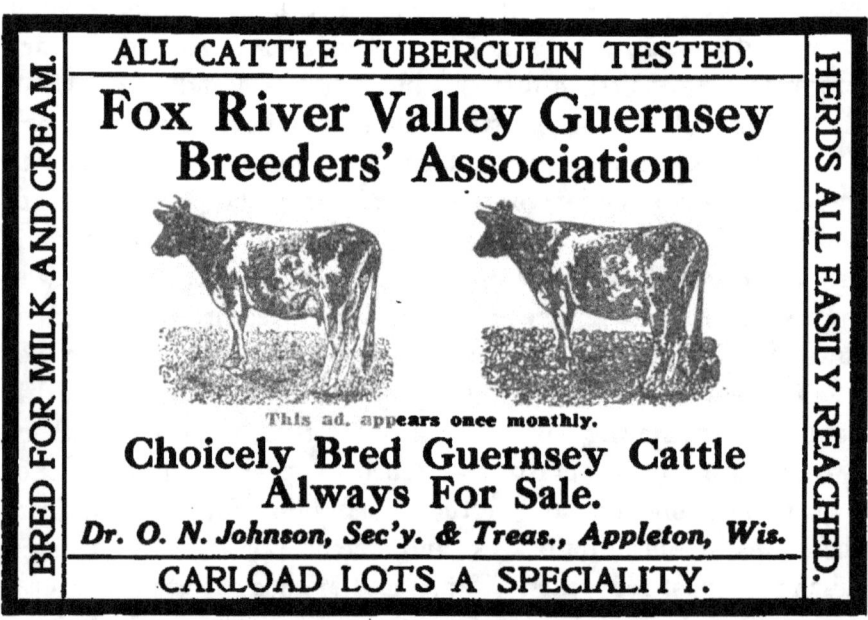

Figure 2. An advertisement carried in leading dairy papers by an active community organization. This advertisement has secured many important orders for the association.

The early activities of these men established a reputation for Lake Mills as being the greatest Holstein center of the Middle West. Buyers have gone there from all parts of the world to buy high-grade and pure-bred Holstein cattle. As high as $175,000 worth of Holstein cattle have been shipped from Lake Mills in a single year. Breeders in the vicinity of Lake Mills have been unable to produce enough dairy cattle to satisfy all of the demand, which was almost an impossibility.

Several of the associations now in existence in Wisconsin are advertising in the leading dairy papers of the country by carrying, at the expense of the associations, such advertisements as are shown in Figures 2, 3 and 4. In reply to inquiries which these advertisements attract, sales lists, published at intervals determined upon by the association, are sent out, such as Figures 5 and 6. These sale lists contain a complete enumeration of all stock for sale together with the names of the respective owners. From this sale list the buyer may choose whatever he may desire.

A community organization may also render great assistance to its members in selling various farm products and in buying feeding stuffs and other supplies. Through the efforts of competent committees and the united action of the members of an association large quantities of a given product can be offered under a specific brand or trade mark. This will insure a uniform quality and rarely ever fails to find a ready market at prices considerably above the average paid for unclassified and ungraded products. Large corporations as well as all large concerns and companies secure advantages in the conduct of their business and usually receive a liberal discount on large quantities of any article they may purchase. There is no reason

---

C. R. Montague, Pres.   Garfield Muckleston, Treas.
E. G. Schley, Vice-Pres.   Dr. David Roberts, Consulting Veterinarian.

WAUKESHA COUNTY HOLSTEIN-FRIESIAN Breeders' Association

Robert L. Baird, Secretary.

Telephone No. 1795.   Waukesha, Wisconsin.

Send for Sale List. Address Secretary for Sale List of Pure Bred and Grades.

Figure 3. Another type of community association advertisement.

Waukesha County Guernsey Association

Animals of both sexes for sale containing some of the best blood of the breed. . .

Herds Tuberculine Tested.

For Sale List Address

F. E. FOX, Secretary.

Figure 4. An association card which is placed in leading dairy papers

---

why a community of farmers cannot receive the same advantage when they become well enough organized to do their business after business-like methods. We have one record on file of how a milk and cream shippers' association (which in reality is a part of one of the associations herein mentioned), maintained a stated price which a dealer tried to cut. He would have succeeded in doing so had not the members of the association firmly opposed him by withholding their product for a short time. The difficulties of successfully organizing communities of farmers for commercial purposes is fully appreciated, but sooner or later the intensive system of farming necessitated by the demand for farm products must bring about such organization.

Figure 5. Sale Lists and Bulletins are issued at regular intervals and distributed to prospective buyers, furnishing a very valuable means of holding the organization together and of advertising its work.

ADVANTAGES IN COMBATING DISEASE. The organized effort of all the people of every community is needed to control tuberculosis. Each community organization can employ a man to test all herds and clean up barns infected with this disease and then insist upon every animal thereafter purchased having a certificate of health. It should be only a matter of a short time when the state will become sufficiently organized in its cattle breeding operations to entirely control this disease. Much credit will be due to local breeders' associations if they will exercise their opportunity and power to this end. Further than combating tuberculosis, members of an association can learn through its educational advantages how to combat other diseases, how to ventilate and provide sunlight in their stables and how to maintain sanitary conditions that will prevent many losses from diseases which are quite common because of a lack of intelligent management on the part of many dairymen.

WHAT COMMUNITY ORGANIZATION DOES NOT DO

It does not compel any man to join with his fellow breeders in the work of breeding and rearing better and more useful dairy cattle. It merely offers him the opportunity.

It does not furnish pure-bred bulls free of charge or insure a herd of high grade or pure-bred dairy cows without effort on the part of the farmer. It simply offers suggestions and encouragement as to how one can accomplish this end.

It does not accomplish results in a single year that will change the type and color of a man's herd unless he has money which will warrant his selling his native cows and replacing them with high grades or pure-breds. It tends, however, to keep a man persistently at the work of improving his cattle until sooner or later he secures a herd of one general type and color.

It does not sell a man's surplus cattle for him nor dictate what prices he shall accept for them. It aids him in coming in contact with the outside world and with buyers whom he must seek to please and make sales as he has opportunity to correspond or come in personal contact with them.

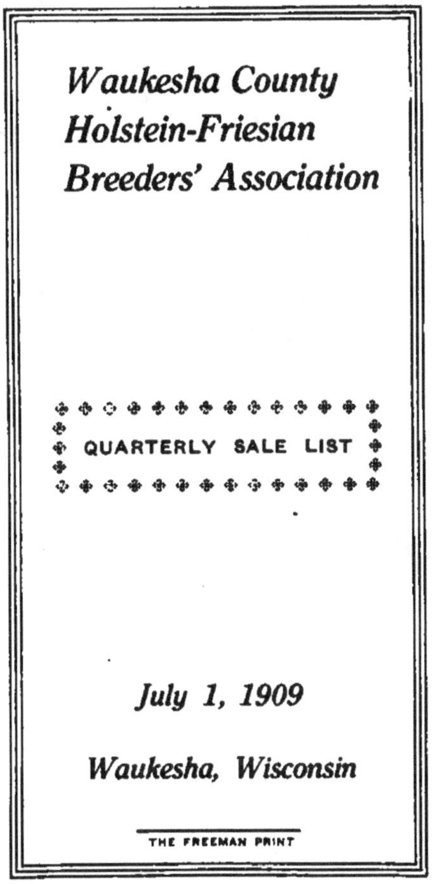

It does not in any way destroy a man's individuality but sharpens his interest and brings out his personality more prominently by placing him in keen competition with his fellow breeders.

Figure 6. Cover of a Sale List published by another association, containing names of breeders with stock for sale.

It does not necessarily cause over production for the reason that no man is obliged to rear any more cattle than he would see fit to keep if he were not a member of a community organization. If community organization encourages a man to keep improved cows in his herd it will certainly have done him a good turn even if it should result in over production.

METHOD OF ORGANIZATION

The plan is that of local organization, as previously stated, and it is therefore within the power of any leading person of a given community to call a meeting for the organization of a cat-

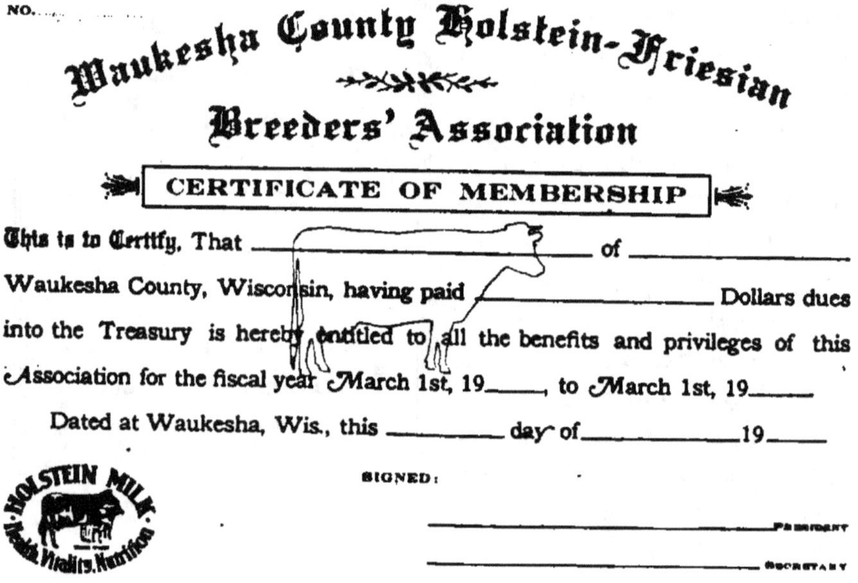

Waukesha County Holstein-Friesian Breeders' Association

CERTIFICATE OF MEMBERSHIP

This is to Certify, That _____ of _____ Waukesha County, Wisconsin, having paid _____ Dollars dues into the Treasury is hereby entitled to all the benefits and privileges of this Association for the fiscal year March 1st, 19____, to March 1st, 19____.

Dated at Waukesha, Wis., this _____ day of _____ 19____.

SIGNED:

_____ PRESIDENT

_____ SECRETARY

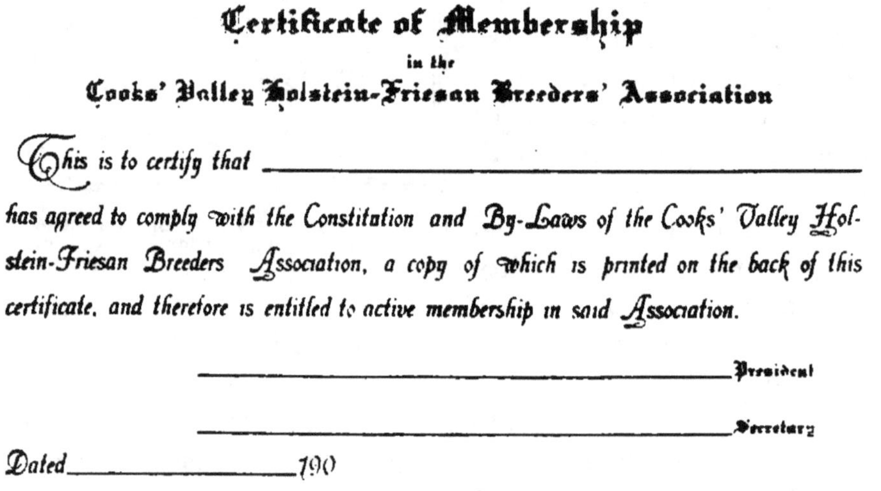

Certificate of Membership

in the

Cooks' Valley Holstein-Friesan Breeders' Association

This is to certify that _____ has agreed to comply with the Constitution and By-Laws of the Cooks' Valley Holstein-Friesan Breeders Association, a copy of which is printed on the back of this certificate, and therefore is entitled to active membership in said Association.

_____ President

_____ Secretary

Dated_____ 190

Figure 7. Two styles of membership certificates used by community organizations. These are printed upon paper of good quality forming a credential which the members may display to indicate their connection with the community association. The certificate shown at the top is printed in three colors.

tle breeders' association. As many people as possible who have a common interest in dairying and in the improvement of a given breed of cattle should be called together. The person or committee who calls the meeting will do well to prepare a program by inviting an outside speaker to deliver an address on the object of the meeting. If a sufficient number of people are interested in the plan to warrant perfecting an association, a constitution and set of by-laws can be adopted, officers elected, and as many members secured as possible to become charter members. Once the organization is perfected its success will depend upon the activity of its members. Cooperation so far as it is possible in aiding and giving encouragement to the work of building up the dairy interests and improving the dairy cattle of the community should be the watchword.

The amount of territory over which an association is to have jurisdiction should not be large, for the reason that if members are too far separated from one another, it is difficult for them to attend meetings and to co-operate with one another. If the interest in the improvement of dairy cattle were as great as it should be, one township would undoubtedly be as much territory as any one association should include. Under present conditions, for each county to have one association for each breed represented, will be most satisfactory. This is the opinion of the majority of breeders who have considered the subject. There need be no difficulty in deciding this question, however, since the object is the organization of dairymen who feel themselves more or less thrown together by one means or another, for the purpose of mutually aiding one another as here outlined.

To have an association for each breed represented in a community or county has the advantage of making it possible for men to discuss topics which bear more specifically on the object of their association. For example, they can very profitably discuss breed interests and points of excellence in their respective animals which it would be more or less difficult to do if the membership were composed of breeders of Holsteins, Jerseys, and Guernseys, each having more or less preference for his particular breed. There is also much better opportunity to manage the sale of stock where each breed is represented by a distinct association. There is room for several associations in each county between which there can be friendly rivalry without harm. Once a year or as often as it seems advisable all associations in a given community can hold joint meetings and discuss topics of general interest and perhaps get the benefit of expert authorities who may be induced to visit the community from time to time.

There should be no limit to the membership as long as men of honor and earnest intentions can be secured as members. Members should always endeavor to keep in good standing with the association and they will soon realize that they will profit by the association to the extent that they share in its work. Membership dues vary as will be noted below and should be in accordance with the amounts necessary to operate the affairs of the association. As soon as an association becomes established and its members have considerable surplus stock to offer, it will be wise to spend more or less money for advertising by some such means as that already referred to on pages 10-13. Live, active meetings with programs which are instructive and entertaining should be provided at such seasons and times as members and their friends and neighbors can attend.

The following Constitution and set of By-Laws are herewith submitted as a suggested basis in perfecting the organization of a breeder's association.

CONSTITUTION.

Article I. Name.

The name of this Association shall be the........................ Breeders' Association.

Article II. Object.

The object of this Association shall be to promote the breeding and improvement of high grade and pure-bred.................cattle in.....................County and to aid its members in buying, breeding and selling first class animals.

Article III. Membership.

The membership shall consist of persons interested in the object of this Association and paying the required annual fee.

Article IV. Organization.

The officers shall be a President, a Vice-President for each township represented, a Secretary and a Treasurer.

There shall be an Executive Committee of five members, which shall have charge of the affairs of the Association when it is not in session, and during its meetings shall be at the command of the Association. This committee shall consist of the President, Secretary, Treasurer and two members elected by the Association at its annual meeting.

ARTICLE V. MEETINGS.

There shall be a regular annual meeting of the Association and such special meetings at times and places determined by the Executive Committee.

ARTICLE VI. ELECTION.

The election of officers shall be held at the regular annual meeting and such election shall be by ballot.

ARTICLE VII. AMENDMENTS.

Amendments to this constitution may be made by a majority of the Executive Committee with the concurrence of two-thirds of the members of the Association voting upon the question by mail within 30 days after the notice is mailed by the Secretary, or by a two-thirds vote of the active members present at the annual meeting.

BY-LAWS.

SECTION 1. NEW MEMBERS.

Any person, upon recommendation of a member and accepted by the Executive Committee, shall become a member upon paying the Secretary the regular annual fee.

SECTION 2. DUTIES AND PRIVILEGES OF MEMBERS.

It shall be the duty of every member to improve his herd of cattle by mating his cows exclusively with pure-bred..................bulls and doing what he can to care for his herd in an up-to-date manner.

It shall also be the duty of members to cooperate so far as possible with their fellow members in the use of pure-bred bulls and in buying and selling animals; also to get new members and encourage them in the practice of better methods in caring for their herds.

All members in good standing shall be entitled to vote in the business meetings of the Association.

SECTION 3. DUES.

The membership dues shall be $............, payable annually to the Secretary of the Association.

SECTION 4. DUES IN ARREARS.

A member in arrears over one year, shall cease to be a member, but may be restored by paying all dues in arrears.

Section 5. Officers.

The officers shall be elected to serve one year and shall perform such services as are ordinarily required by their positions and shall serve until the election of their successors.

Section 6. President.

The President shall serve for one year and shall preside over the meetings of the Association and shall give an annual address.

Section 7. Vice-President.

It shall be the duty of the several Vice-Presidents to look after the interests of the Association in the various townships and they shall have the privilege of calling local meetings and doing all in their power to promote the general interests of the Association in such manner as the Executive Committee shall deem fit.

Section 8. Treasurer.

The Treasurer shall receive and hold all funds coming to the Association, and shall disburse or invest such money as directed by the Executive Committee and shall keep an accurate and detailed account of all receipts and disbursements and make a report of the same to the Executive Committee and to the Association at each annual meeting. The records and accounts of the Treasurer shall be open to the inspection of the members. The Executive Committee may require a suitable bond of the Treasurer whenever in their judgment it appears advisable to do so.

Section 9. Secretary.

The Secretary shall keep a record of all proceedings of the Association and of the Executive Committee, all membership dues and miscellaneous receipts, and pay all moneys received by the Association promptly to the Treasurer. He shall send and receive all notices and record and hold in trust such property of the Association other than money in the hands of the Treasurer. He shall also act as correspondent for the Association in such matters as pertain to the business of the Association and do all in his power to promote the interests of the Association.

Section 10. Executive Committee.

The President shall act as chairman of the Executive Committee and the meetings shall be called through the Secretary. Three of the five members shall constitute a quorum. It shall be the duty of this committee to determine upon the place and time of the annual and special meetings and give due notice of them through the Secretary. They shall elect members of the Association and shall have power to expel any member whenever in their judgment it is for the best interests of the Association to do so. They shall carry out the resolutions voted by the Association, appoint such special committees as necessary and make an annual report to the Association upon the standing and progress of the work of the Association.

Section 11. Auditing Committee.

At each annual meeting there shall be elected an auditing committee consisting of three members, whose duty it shall be to examine and report upon all books and accounts of the officers for the ensuing year.

Section 12. Order of Business.

1. Reading of minutes of previous meeting.
2. Report of Treasurer.
3. Reports of committees.
4. Unfinished business.
5. New business.
6. Election of officers.

Section 13. Rules of Order.

The meeting of the Association shall be governed by Robert's Rules of Order.

Associations in Wisconsin

The following local associations were organized in Wisconsin before January 15, 1910. In the majority of cases aid has been given by the Agricultural College in the matter of establishing these associations, speakers having been sent to address meetings that have been held for this purpose. It will be noticed that with few exceptions these associations represent specific breed interests. Those of a general character are the result of a desire on the part of dairymen to improve their cattle and agricultural conditions without giving special emphasis to any one breed. Data are given with reference to the names of the associations and their secretaries, the dates of organization, the number of charter and present members, the annual membership dues, and such other items of general interest which have been furnished. The associations are mentioned in the order of the dates they were organized.

TABLE I.—LIST OF COMMUNITY BREEDERS' ORGANIZATIONS IN WISCONSIN—JANUARY, 1910.

Name of Association.	Date of organization.	Membership. Charter.	Membership. Present.	Annual dues.	Increase in the number of pure-bred sires in service since time of organization.	Approximate number of pure bred animals owned by members.	Remarks.
Waukesha Guernsey Breeders'. Secy. F. E. Fox, Waukesha.	Spring, 1906	10	65	$1.00	50	1000	A spring and summer meeting is held each year. Sale list published quarterly. Advertisements carried in dairy papers.
Barron County Holstein Breeders'. Secy. Charles Stair, Barron.	Jan. 15, 1907	63	98	1.00	90%	75	Effort is put forth to have all herds tuberculin tested.
Clark County Holstein Breeders'. Secy. Geo. L. Jacques, Neillsville.	Jan. 26, 1907	23	45	.50	25	29	
Clark County Guernsey Breeders'. Secy. L. Williamson, Neillsville.	Feb. 12, 1907	12	20	.50	50%	15	Majority of herds tuberculin tested annually.
Barron County Guernsey Breeders'. Secy. Jacob Kohlen, Barron.	Feb. 13, 1907	21	68	.50	12	25	Effort being made to have herds tuberculin tested annually.
Cook's Valley Holstein-Friesian Breeders'. Secy. J. Wesley Raven, Bloomer.	Mar. 4, 1907	7	20	1.00	9		Meetings held quarterly.
LaCrosse County Guernsey Breeders'. Secy. H. W. Griswold, West Salem.	Mar. 20, 1907	28	35	.50	10	130	Meetings held annually. 3600 head tuberculin tested
Westfield Farmers' Club. Secy. R. G. Nieman, Loganville.	Mar. 20, 1907	45	45	.50	10	40	Monthly meetings held. Tuberculin tests have been applied in several instances
Barron County Jersey Breeders'. Secy. E. C. McClelland, Rice Lake.	March, 1907			.50			
Milltown Agricultural Society. Acting Secy. Geo. A. Nelson Milltown.	June 4, 1907	26	23	.50		28 cattle & 45 hogs.	
Monroe County Jersey Breeders'. Secy. W. T. Aney, Norwalk.	June, 1907	15	15	.50		50	All herds tuberculin tested.
Marshfield Guernsey Breeders'. Secy. W. E. Hargrave, Marshfield.	June, 1907	18	27	.50	6	18	
Dunn County Dairymen's and Breeders'. Acting Secy. F. C. Jacobs, Elk Mound.	Dec. 20, 1907	16	64	.50* 1.00	15	125	
Price County Holstein Breeders'. Secy. C. W. Welch, Prentice.	Jan. 7, 1910	30	30	1.00			
Price County Guernsey Breeders'. Secy. J. S. Clark, Prentice.	Jan. 7, 1910	20	20	1.00			

COMMUNITY BREEDERS' ASSOCIATIONS.

Association and Secretary	Date			Fee			Remarks
Central Wisconsin Holstein-Friesian Breeders'. Secy. A. E. Howard, Marshfield	Mar. 20, 1908	28	60	.25 $1.00*	12	450	Several members applying tuberculin test annually.
Guernsey Breeders' Ass'n of Jackson County. Secy. F. A. Sly, Sechlerville	Mar. 27, 1908	6	16	1.00	10	25	
Fox River Valley Guernsey Breeders'. Secy. Dr. O. N. Johnson, Appleton	Mar. 1908	13	13	3.00	11	72	All herds tuberculin tested annually. Advertisements carried in dairy papers.
Fond du Lac County Guernsey Breeders'. Secy. A. F. Hintz, Ripon	Sept. 3, 1908	9	20	.50	24	250	Most herds tuberculin test'd.
Rusk and Chippewa Co's Guernsey Breeders'. Secy. Wm. Grand, Jim Falls	Sept. 4, 1908	13	13	.50	4	30	
Southwestern Guernsey Breeders'. Secy. Charles Wilkins, Livingston	Dec. 30, 1908	7	18	1.00	6	30	Many members applying tuberculin test.
Monroe County Guernsey Breeders'. Secy. H. P. Howell, Sparta	Jan. 7, 1909	24	67	.50 1.00*	12	50	Association plans to tuberculin test all herds.
Sheboygan Co. Holstein-Friesian Breeders'. Secy. T. H. Thackray, Glenbeulah	Jan. 16, 1909	20	38	.50		500	Some members have tuberculin tested their herds.
The Kiel Agricultural and Breeders'. Secy. Aug. F. Luedke, Kiel	Jan. 16, 1909	7	54	.50	100%	13	Effort is made to keep out tuberculous animals.
Kewaunee Co. Holstein-Friesian Breeders'. Secy. Wm. Katel, Kewaunee	Feb. 6, 1909	9	11	1.00	4	4	
Waukesha Co. Holstein-Friesian Breeders'. Secy. Rob. L. Baird, Waukesha	Mar. 1, 1909	25	74	2.00		500	Effort being made to stamp out tuberculosis. Sale list published quarterly. Advertisements carried in dairy papers.
Sheboygan County Guernsey Breeders'. Secy. S. A. Eastman, Sheboygan Falls	April 2, 1909	7	7	.50		42	6 sires in service.
La Crosse River Valley Holstein Breeders'. Secy. W. H. Ascott, Sparta	May 15, 1909	17	20	1.00			20 pure-herd sires in service owned by members.
Marathon County Guernsey Breeders'. Secy. J. J. Bean, Wausau	Sept. 1909	10	30	1.00		125	20 pure-bred sires in service owned by members.
Green County Holstein-Friesian Breeders'. Secy. J. C. Penn, Monroe	Oct. 30, 1909	17	17	1.00		600	About 25 per cent of herds tuberculin tested.
Fort Atkinson Guernsey Breeders'. Secy. W. J. Heid, Fort Atkinson	Dec. 15, 1909	17	17	1.00			

*Initiation fee.

THE UNIVERSITY OF WISCONSIN

Agricultural Experiment Station

STATION STAFF

THE PRESIDENT of the University
H. L. RUSSELL, Director

S. M. BABCOCK, Assistant Director
IDA HERFURTH, Executive Clerk

W. A. HENRY, Emeritus Professor of Agriculture

A. S. ALEXANDER, Veterinary Science, In charge of Stallion Licensing
S. M. BABCOCK, In charge of Agricultural Chemistry
L. J. COLE, In charge of Experimental Breeding.
E. J. DELWICHE, Supt. Northern Sub-Stations, (Ashland, Wis.)
E. H. FARRINGTON, In charge of Dairy Husbandry
J. G. FULLER, Animal Husbandry
J. G. HALPIN, In charge of Poultry Husbandry
E. B. HART, Agricultural Chemistry
E. G. HASTINGS, Agricultural Bacteriology
K. L. HATCH, Agricultural Education; Secretary Agricultural Extension
G. C. HUMPHREY, In charge of Animal Husbandry
L. R. JONES, In charge of Plant Pathology
E. R. JONES, Soils
C. E. LEE, Dairying

ABBY L. MARLATT, In charge of Home Economics
E. V. MCCOLLUM, Agricultural Chemistry
J. G. MOORE, In charge of Horticulture (Pro tem.)
R. A. MOORE, In charge of Agronomy
C. P. NORGORD, Agronomy
C. A. OCOCK, In charge of Agricultural Engineering
D. H. OTIS, Farm Management
M. P. RAVENEL, In charge of Agricultural Bacteriology
J. L. SAMMIS, Dairy Husbandry
J. G. SANDERS, In charge of Economic Entomology.
JOHN SPENCER, Veterinary Science.
C. W. STODDART, Soils
H. C. TAYLOR, In charge of Agricultural Economics
A. R. WHITSON, In charge of Soils
F. W. WOLL, In charge of Feed and Fertilizer Inspection; Dairy Tests

G. H. BENKENDORF, Dairy Husbandry
EMILY BRESEE, Feed and Fertilizer Inspection
L. R. DAVIES, Dairy Tests; Feed and Fertilizer Inspection
E. E. ELDREDGE, Agricultural Bacteriology
C. S. HEAN, Agricultural Library
LEONA HOPE, Home Economics
J. JOHNSON, Horticulture
J. C. JURRJENS, Feed and Fertilizer Inspection
F. KLEINHEINZ, Animal Husbandry
ALICE LOOMIS, Home Economics
O. G. MALDE, Cranberry Investigations, (Grand Rapids, Wis.)
J. C. MARQUIS, Agricultural Editor
J. G. MILWARD, Horticulture
W. E. MORRIS, Feed and Fertilizer Inspection

J. M. NAPIER, Agronomy
C. R. ORTON, Plant Pathology.
P. P. PETERSON, Soils
W. H. PETERSON, Agricultural Chemistry
A. J. ROGERS, JR., Horticulture; In charge of Nursery Inspection
F. J. SIEVERS, Soils
H. STEENBOCK, Agricultural Chemistry
W. W. SYLVESTER, Agricultural Engineering
A. L. STONE, Agronomy; In charge of Seed Inspection
J. L. TORMEY, Animal Husbandry
W. E. TOTTINGHAM, Agricultural Chemistry
E. TRUOG, Soils
H. L. WALSTER, Soils
W. W. WEIR, Soils
F. WHITE, Agricultural Engineering
W. H. WRIGHT, Agricultural Bacteriology

FARMERS' INSTITUTES

GEORGE MCKERROW, Superintendent
NELLIE E. GRIFFITHS, Clerk

The bulletins of this Station are sent free to residents of the State. Names will be entered on the regular mailing list upon request.

www.ingramcontent.com/pod-product-compliance
Lightning Source LLC
Chambersburg PA
CBHW062345220526
45469CB00008B/2843